MW01129599

How to Explain CLIMATE SCIENCE to a Grown-Up

RUTH SPIRO

Illustrated by TERESA MARTÍNEZ

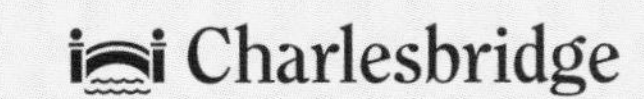

When I was a little kid,

I asked a LOT of questions:

Why does a cupcake taste sweet?

Why does rain fall down?

I thought my grown-up had all the answers.

But now that I'm big, I know the truth.
MY GROWN-UP DOES NOT KNOW ALL THE ANSWERS!
Sometimes they need ME to explain things to THEM.
If you're reading this, I'm guessing YOUR grown-up needs help understanding stuff, too.

With a little help from this book,
YOU can explain CLIMATE SCIENCE to your grown-up!

First, show your grown-up this book.

They may ask,
What is climate science, and why do I need to know about it?
It looks like you have a LOT of explaining to do.
Start with something simple, like . . .

that tree.
That's our favorite!
It usually blooms around my birthday.

Your grown-up has noticed something important. Many trees and other plants are blooming earlier in the spring. This is because Earth's climate is getting warmer.

Pro Tip

Your grown-up may find this hard to believe. Tell them that scientists always provide facts to support what they say. Taking those photos at the same time each year is a way of collecting information—just like scientists do!

Some grown-ups may think that weather is the same as climate, but it's not. Here's how to explain the difference:

Pro Tip

Sometimes scientific words can be confusing. Make sure your grown-up understands them before moving on.

Weather is what the conditions outside are like at a specific time and place. This includes temperature, wind, and precipitation such as rain, snow, or hail.

Climate is the overall pattern of weather over a long time.

Some places are warm and sunny. Others are cold and snowy.

Scientists collect and study information about climate in different places on Earth.

They may use satellite images to observe the size of glaciers or special instruments to measure air, land, and ocean temperatures. They also use computers to create models that predict what could happen in the future.

This observation,
measurement, and modeling
is climate science.

Scientists have discovered that Earth's climate is changing. Why?

Show your grown-up this helpful picture:

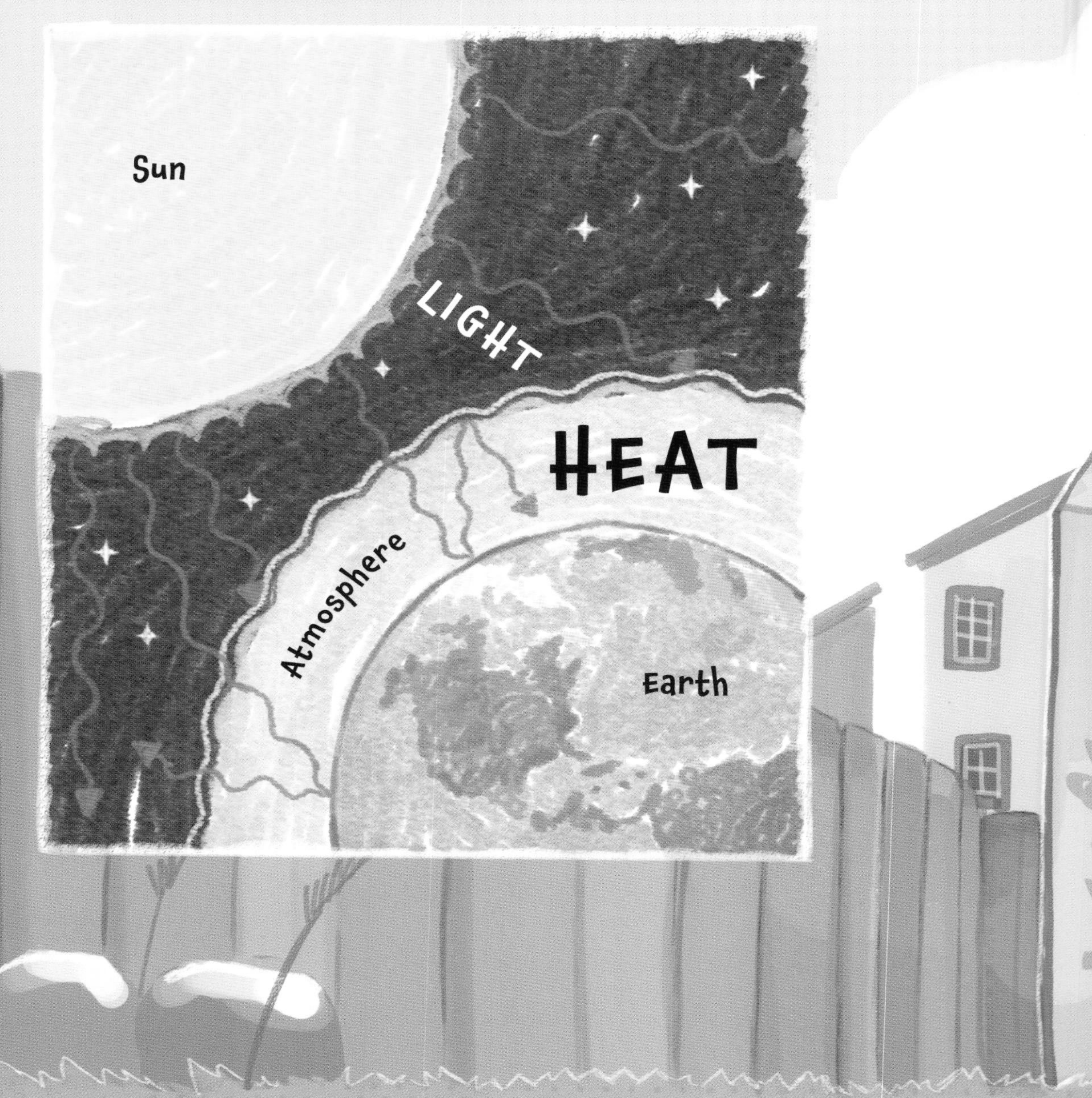

Light from the sun warms Earth. Some of that warmth escapes into space, but some is trapped by gases in the atmosphere.

Tell your grown-up these gases are called greenhouse gases. Just like a greenhouse, they let in light and trap heat. Without them, Earth would be too cold for living things to survive.

Pro Tip

When explaining something new to a grown-up, it helps to compare the new thing to something familiar.

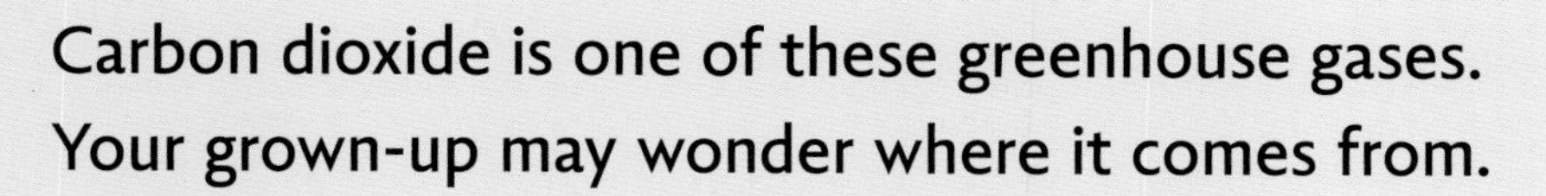

Carbon dioxide is one of these greenhouse gases. Your grown-up may wonder where it comes from.

Carbon is found in all living things on Earth. Over millions of years, plants and animals buried in the ground transform into coal, oil, and gas. These are called fossil fuels.

When fossil fuels are burned to create energy, their carbon combines with oxygen to make carbon dioxide.

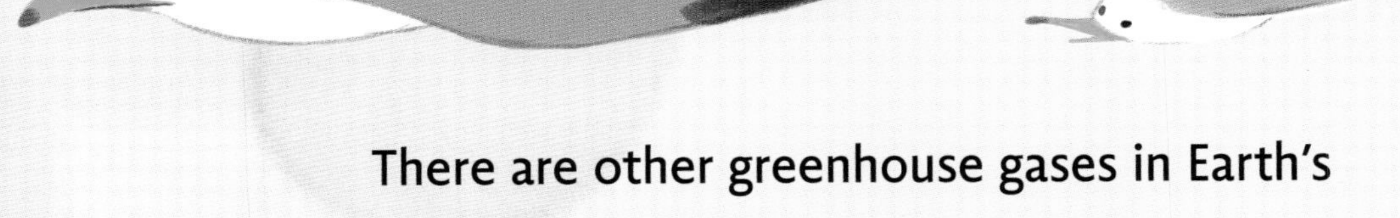

There are other greenhouse gases in Earth's atmosphere, too.

Methane comes from food and plastic breaking down in landfills, and also . . . COW BURPS!

Nitrous oxide comes from fertilizer and also from burning fossil fuels.

Fluorinated gases come from using air-conditioning, refrigerators, and freezers.

For the past ten thousand years, greenhouse gases in the atmosphere have kept Earth's temperature in balance. Not too cold. Not too warm. Just right.

But now humans are tipping the balance.

Burning fossil fuel releases huge amounts of carbon dioxide into the atmosphere. Making and using plastic creates carbon dioxide, too.

Trees help by taking in carbon dioxide. But people are burning and cutting down forests to make room for building and farming.

All this adds up to more greenhouse gases in the atmosphere. Earth is getting warmer, and that's causing the climate to change.

If your grown-up is ready to see some real-world effects of climate change, it's time for a field trip . . .

to the beach!

Your grown-up may notice the water is higher than before—and not just because of the tide.

Explain that the ocean is getting warmer. As the warm water expands, sea levels rise. Glaciers are also melting, adding extra water to the ocean.

Rising sea levels erode beaches and can cause flooding where people and animals live.

Pro Tip
If your grown-up seems worried, a little humor may help.
How can you tell that the ocean is friendly?
How?
It waves at everyone!

You and your grown-up may notice other changes.
Our fishing pier was no match for last week's big storm.
Looks like they're rebuilding it.

With climate change, warmer temperatures can increase the amount of water vapor in the air. This can lead to heavier rains and stronger hurricanes in some areas. Other places may see heat waves or drought.

Pro Tip

Learning about climate change can sometimes be a little scary. Reassure your grown-up that you will always help keep each other safe!

Your grown-up may be wondering how to solve such a big problem. First, reassure them there's good news.

It's true that Earth's climate is warming because of humans. But humans can also make changes that will repair the planet.

Pro Tip

Big problems can feel frustrating. Remind your grown-up that it's okay to talk about it with someone.

Tell your grown-up the most important goal:

Reduce greenhouse gases!

Greenhouse gases come from things we do every day:

- Burning fossil fuel when we drive a gas-powered car
- Using electricity that comes from burning fossil fuel
- Adding food and plastic waste to landfills

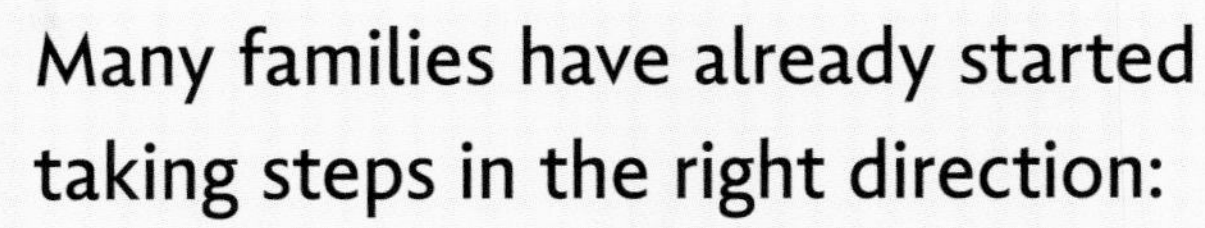

Many families have already started taking steps in the right direction:

- Walking or biking instead of driving
- Turning off lights when not in use
- Composting food waste and buying less plastic

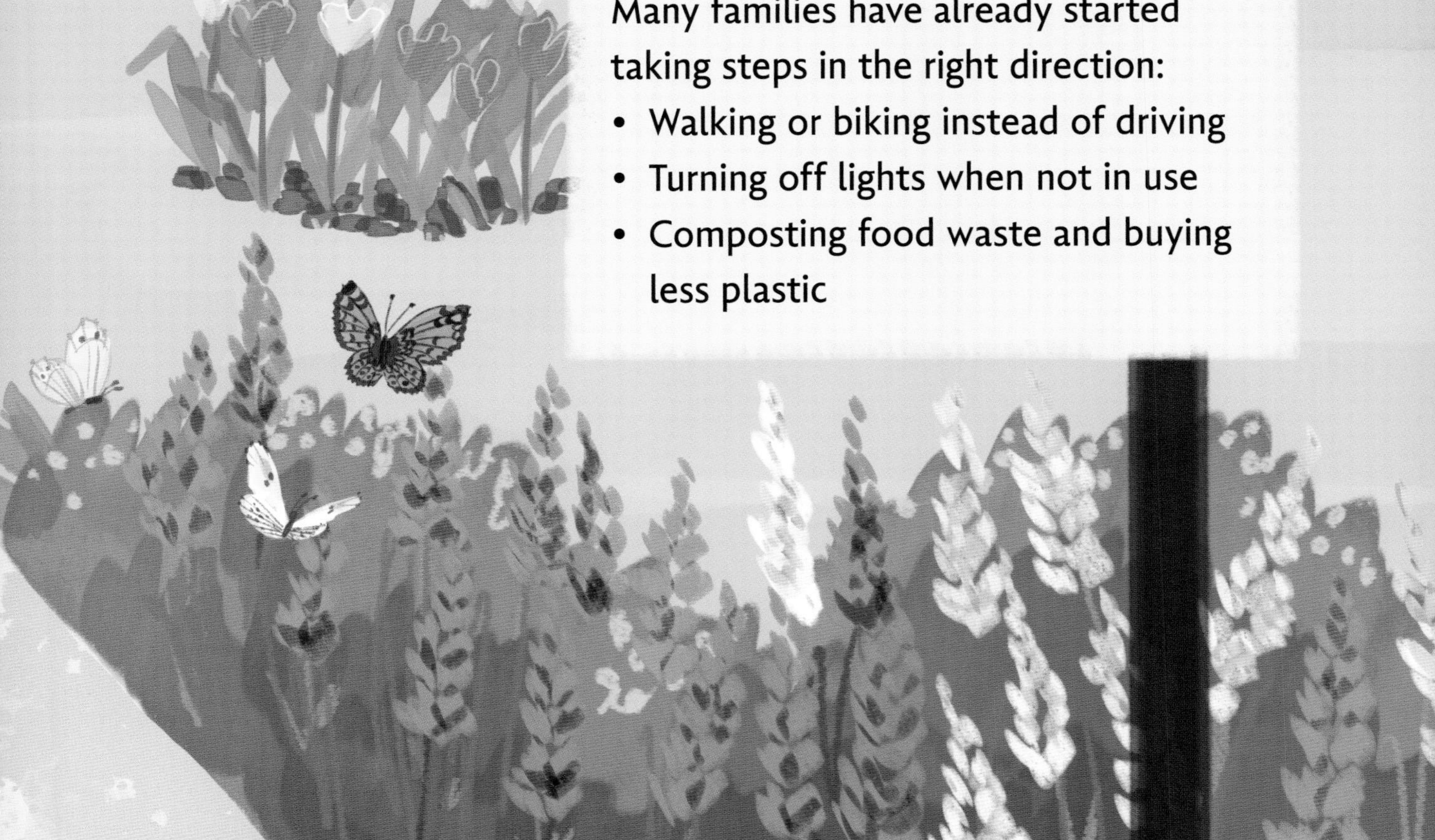

Making changes at home is a good start, but the planet needs humans to GO BIG! Help your grown-up come up with ways to make a big difference.

Using less electricity is good . . .

but it's even better if you ask your local government to switch your community to clean, renewable energy sources like solar and wind.

It's great if you compost at home . . .

but it's even better if you can convince your entire school district to compost food waste from student lunches.

Pro Tip

Encouraging your grown-up to take action will help them see that they have the power to make positive changes in the world.

Reducing your own plastic use is always a good idea . . .
and so is writing letters to companies asking them to use less plastic in their packaging.

You've explained a LOT about climate science to your grown-up. Now is a good time to make your own Climate Action Plan together.

Ask your grown-up:

What signs of climate change did you notice today?

What's causing these problems?

What are some things we can do at home, starting today?

Finally, make a list of WHY humans need to do their part to help the Earth. Put it somewhere you'll see it every day.

Congratulations!

YOU DID IT!
You explained climate science to a grown-up!
Get ready, because now your grown-up
is going to have a LOT more questions.

What will you explain next?

Glossary

atmosphere: The layer of air surrounding Earth. The atmosphere includes the air we breathe as well as gases that help keep Earth warm.

climate: The overall pattern of weather in a place over a long time (thirty years or more).

climate change: A change in temperature and weather patterns observed over a long period of time. Modern climate change is mainly caused by humans adding more greenhouse gases to the atmosphere.

climate science: The study of Earth's climate, how it is changing, and how it affects humans and other living things.

fossil fuel: Coal, oil, or gas that comes from plants and animals buried in the ground for millions of years. Burning fossil fuels to create energy adds more carbon dioxide to the atmosphere.

greenhouse gas: A gas in Earth's atmosphere that lets in sunlight and keeps heat from escaping into space. Greenhouse gases include carbon dioxide, methane, nitrous oxide, fluorinated gases, and water vapor.

model: In science, a complex math equation used to make sense of large amounts of information. Climate scientists use computer models to explore how different amounts of greenhouse gases affect Earth.

renewable energy: Energy that comes from natural sources that do not run out, such as sunlight, wind, and water. Also called green energy or clean energy.

weather: The outdoor conditions in a place at a specific time. This includes temperature, wind, and precipitation like rain, sleet, snow, or hail. The weather can change throughout the day.

Ways you and your grown-up can take action TODAY!

At Home

- Turn off lights when not in use.
- Use air-conditioning only when necessary.
- Use less plastic.
- Walk or bike instead of driving, if possible.
- Join a citizen-science project to help scientists gather data related to climate.

At School

- Encourage classmates to reduce waste at lunchtime.
- Compost food and yard waste.
- Ask grown-ups to turn off cars at pickup rather than keeping engines running.
- Plant trees, native plants, and school gardens!

In the Community

- Start or join a community garden.
- Ask local government to plant more trees.
- Ask community leaders to consider installing roundabouts at road intersections.
- Look for ways to reduce waste. You could join a local "buy nothing" group or organize a "repair café."
- Ask your public library to set up a display of books about climate science—including this one!

Most important: Speak up and write letters to those in power, including elected officials. Explaining climate change to grown-ups can lead to big changes!

This book is dedicated to YOU, and to everyone working to protect our planet, in ways both big and small—R. S.

To Kaïna and her orchids—T. M.

Very special thanks to Andrew Pershing, VP for Science at Climate Central, for sharing his invaluable expertise and advice.

Published by Charlesbridge
9 Galen Street, Watertown, MA 02472
(617) 926-0329 • www.charlesbridge.com

Printed in China
(hc) 10 9 8 7 6 5 4 3 2 1

Illustrations created digitally using Photoshop and a Wacom tablet
Display type set in Chaloops by The Chank Company
Text type set in Fontanella by Guisela Mendoza
Printed by 1010 Printing International Limited in Huizhou, Guangdong, China
Production supervision by Mira Kennedy
Designed by Cathleen Schaad

Library of Congress Cataloging-in-Publication Data
Names: Spiro, Ruth, author. | Martínez, Teresa, 1980– illustrator.
Title: How to explain climate science to a grown-up / Ruth Spiro; illustrated by Teresa Martínez.
Description: Watertown, MA: Charlesbridge, [2025] | Series: How to explain science to a grown-up | Audience: Ages 4–8 | Audience: Grades 2–3 | Summary: "In this tongue-in-cheek guide, a kid expert explains to young readers how to teach their grown-up about the basics of climate science and global warming."—Provided by publisher.
Identifiers: LCCN 2023056151 (print) | LCCN 2023056152 (ebook) | ISBN 9781623546205 (hardcover) | ISBN 9781632892652 (ebook)
Subjects: LCSH: Climatic changes—Juvenile literature.
Classification: LCC QC903.15 .S68 2025 (print) | LCC QC903.15 (ebook) | DDC 363.7—dc23/eng/20240324
LC record available at https://lccn.loc.gov/2023056151
LC ebook record available at https://lccn.loc.gov/2023056152